CELESTE
TUNNELS
UNDERGROUND

Written by Courtney Kelly, P.E.

Illustrated by Erin Nielson

Copyright © 2024 Courtney Kelly

All rights reserved. No part of this book may be reproduced in any form or by any electronic or mechanical means, including information storage and retrieval systems, without permission in writing from the publisher, except by reviewers, who may quote brief passages in a review.

www.courtneykellybooks.com

ISBN 979-8-9903082-1-3 (Hardcover)
ISBN 979-8-9903082-2-0 (Paperback)
ISBN 979-8-9903082-0-6 (E-Pub)
Library of Congress Control Number 2024910343

Some events and places in this book are nonfictional. Readers are encouraged to research and learn more about them.

Edited by: Courtney Kelly
Cover, illustrations, and interior layout by: Erin Nielson

Courtney Kelly Books
Printed in the United States of America

Discount pricing is available for bulk orders (25 or more books). For more information, please contact us at info@courtneykellybooks.com

This is the second book of the *Celeste Saves the City* collection. To find out more about the first book, visit www.courtneykellybooks.com or search for it anywhere books are sold.

This book is dedicated to all of the women who work in the architecture, engineering, and construction industry. May your brilliance shine everywhere you go!

- Courtney Kelly

You may have heard of Celeste, who saved the city of New Orleans by protecting its coast with barrier islands built by her amazing team.

The wetlands are growing and getting stronger every day.
Now she has a problem to solve in Dallas without delay!

A creative solution is needed to open up more space.

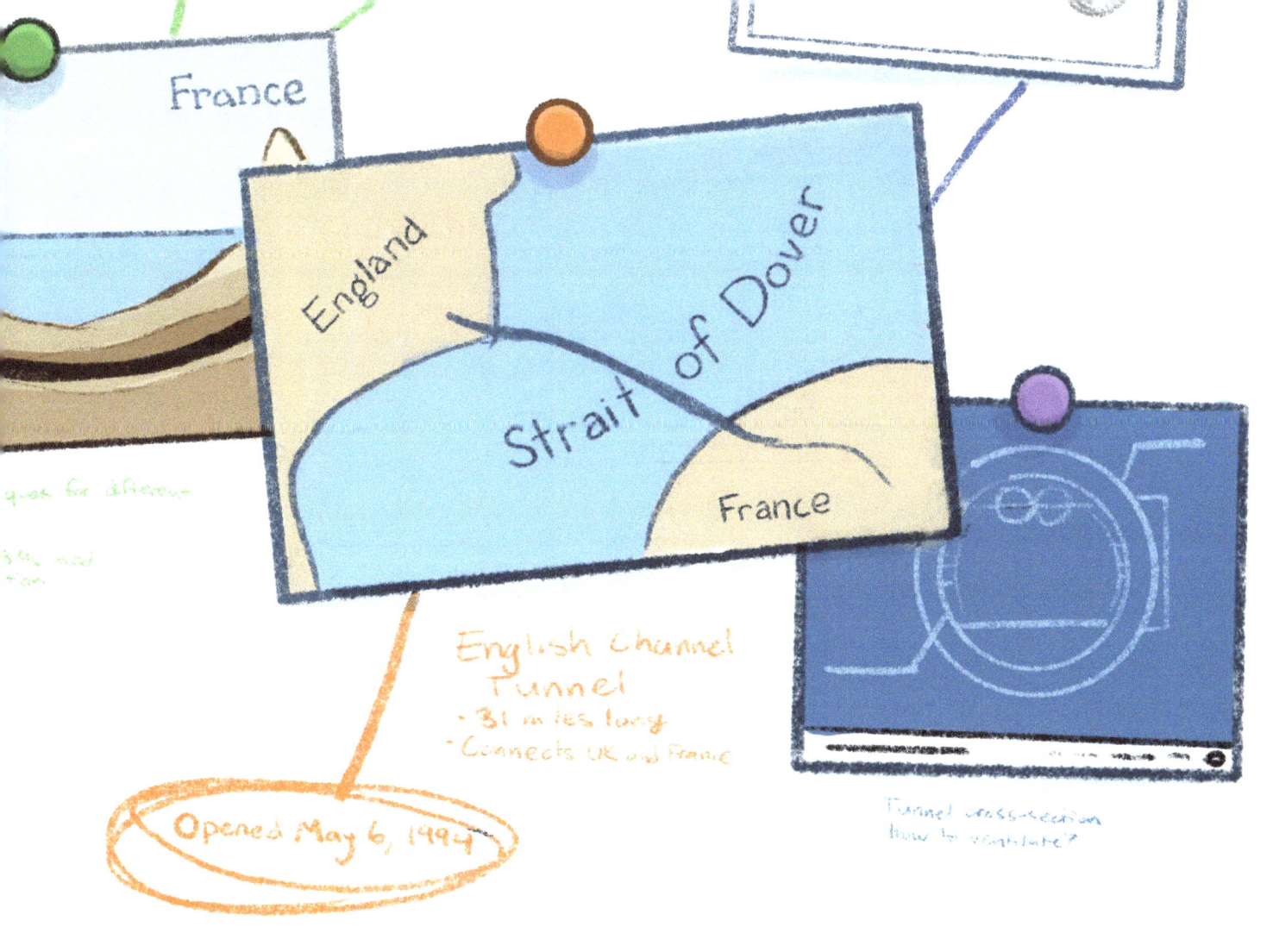

Finding room above is one way to go,
but Celeste ponders what it would be like to put things below.

She found out about tunnels, like the English Channel, that run deep underground
and help to keep people and other things moving around.

To the library she went on a mission to learn more.
With shelves filled to the brim with books, there was so much to search for.

Her studies revealed that tunnels would be great in many ways,
like reducing noise and helping the air stay clean and safe!

Celeste took off to explore how tunnels could help fill the need and ease congestion on Dallas' busy streets.

She traveled far and wide to experience how tunnels were used. What she found opened her eyes to the amazing things they could do!

In Boston, the Big Dig brought pollution levels down
and made space for the Rose Fitzgerald Greenway downtown!

Between Denmark and Germany, the world's longest underwater rail and road tunnel was being built under the Baltic Sea. When finished, people will be able to travel from one country to the next with ease.

Freight trains barreled deep under the Swiss Alps in fact.
Named the Gotthard Base Tunnel, it's the world's deepest at that!

The Guanajuato tunnels in Mexico City were created to keep water from flooding the town, but they are now used as roads for people to walk and drive around.

After seeing the great projects done by other civil engineers,
Celeste was confident that she would be able to come up with a solution here.

Together with a new team who are experts at building underground, they started on a mission to dig deep under the town.

Using a bore machine that was as big as Big Tex, slowly, but surely, they inched along from start to end after the path was checked.

Mile by mile, dirt was moved out of the way as the drilling head cut through the earth, excavating the dense clay.

Once the opening was made, it was time to finish the Achiever Tunnel off.
In came tradeswomen and tradesmen ready to work on this important cause.

Carpenters, masons, operators, and electricians took to their crafts like pros,

making the space safe for people to drive in when they are on the go.

It took some time, but the work was finally through, and Celeste welcomed the first guests as the city celebrated this breakthrough.

With this challenging task now complete,
Celeste admired her great feat!
Where she ends up next you'll have to wait and see,
but there is sure to be another place in need.

About the Author

Born and raised in New Orleans, Louisiana, Courtney grew up wanting to be a veterinarian - until Hurricane Katrina changed the course of her life. After attending numerous engineering and math camps during high school, she left Louisiana to pursue degrees in civil engineering and mathematics at Southern Methodist University (SMU) in Dallas, Texas. Courtney went on to obtain a master's degree in civil engineering with a concentration in structures from SMU and an MBA from Lamar University in Beaumont, Texas. Her career in construction project management of commercial and heavy civil infrastructure projects has included work at DFW International and Love Field Airports, highways, and local municipalities. In 2023, she was selected as an Engineering News-Record Texas & Louisiana Top Young Professional. She is also a licensed professional engineer in the state of Texas. In her spare time, she enjoys traveling, attending arts events, discovering new music, watching Asian dramas, and spending time with her fluffy bunny, Albus.

About the Illustrator

Erin Nielson is a freelance illustrator in McKinney, Texas. She loves to create imaginative and whimsical artwork, especially if that artwork can include a few animals. She has illustrated several picture books, including *Celeste Saves the City* and *'Twas the Day Before Christmas in Bethlehem Town*. Erin's love of art started early, with her paintings and artwork hung in her grandmother's home. She attended college at Brigham Young University, and graduated with a BFA in illustration. Soon after graduating, she married her husband, Matt, and moved to Texas for new adventures. She now enjoys working out of her home studio with her two kids and her rescue dog Zoey.

Milton Keynes UK
Ingram Content Group UK Ltd.
UKHW021147260624
444643UK00005B/21